GERMAN
LIGHT AND HEAVY
INFANTRY ARTILLERY
1914 - 1945

At a full gallop, a light infantry gun follows the advancing infantry. Four members of the gun crew are seated on the limber and ammunition vehicle for the light infantry gun (Inf. 14). This photo was taken during prewar maneuvers.

Wolfgang Fleischer

Schiffer Military/Aviation History
Atglen, PA

ACKNOWLEDGEMENTS

The author would like to thank Frau Sonja Wetzig (Research), Herr Günther Thiede (Photography) and Herr Dirk Hensel (Ammunition Drawings). Thanks are also due to Herr K. H. Caye of the Military History Collection in Kiel-Holtenau and Herr Georg Driess of the Museum for Handicrafts, Nature and Weapons Technology in Dahn-Reichenbach, as well as Herr Dieter Heinrich of the Defensive Technological Study Collection in Koblenz.

BIBLIOGRAPHY

Among others, H.Dv. 105/2 "The Mine Launcher", H.Dv. 111/2 "Firing Instructions for Infantry Guns", H.Dv. 119/291 "Shot Tables for the Light Infantry Gun 18 . . .", D 481/8 "Manual for the Ammunition of the Heavy Infantry Gun 33, . . .", D 2025 "The Heavy Infantry Gun 33/1", MZA Potsdam WF 10/2460. WF 10/13551 and WF 10/13675, Muther, A., "The Equipment of the Light Artillery before, in and after the World War, Part II: Infantry Guns, Antitank Defenses and Tank Equipment", Military Weekly News, Military Scientific Announcements, Defense Technology Monthly; Signal, The Wehrmacht.

Equipping the infantry with the heavy infantry gun was a German peculiarity. In no other army were 15-centimeter guns included organically on the level of infantry regiments.

Translated from the German by Edward Force

Printed in the United States of America.
ISBN: 0-88740-815-X

This title was originally published under the title,
Waffen-Arsenal Waffen und Fahrzeuge der Heere und Luftstreitkräfte Deutsche Leichte und Schwere Infanteriegeschütze 1914-1945,
by Podzun-Pallas-Verlag, Friedberg.

PHOTO CREDITS

Caye (3), Driess (2), Kuntzsch (1), Fleischer (68), Federal Archives, Koblenz (1), Defense Technological Study Collection, Koblenz (4), Military History Museum, Dresden (2).

Published by Schiffer Publishing Ltd.
77 Lower Valley Road
Atglen, PA 19310
Please write for a free catalog.
This book may be purchased from the publisher.
Please include $2.95 postage.
Try your bookstore first.

We are interested in hearing from authors
with book ideas on related topics.

The 7.7 cm Field Cannon 96 n.A. ranked among the lightweight field cannons of its day. Weighing 925 kilograms in firing position, though, it was too heavy in terms of what was needed for an infantry escort gun. This photo was taken in 1914.

The origins of the development of German infantry guns may be found in World War I. At the beginning of the war, the field artillery of the German Army had two types of guns: the light 7.7 cm Field Cannon 96 n.A., and the 10.5 cm Field Howitzer 98/09. There was a deep-seated conviction that these two weapons would be able to fulfill all required tasks. This faith, though, could not be confirmed, as during the course of the war it turned out that the infantry needed particularly designed guns with very specific performance parameters in order to overcome strong opposition.

What reasons now forced the abandonment of the prewar viewpoints?

1. The upgraded effect of weapons required placement of the field guns far behind the foremost infantry lines. Firing from concealed positions became the rule.

2. Cooperation between the field artillery and the infantry was made more difficult by the greater distances. Thus many small targets, scattered about the battlefield, could withstand heavy fire and seriously hinder the following infantry attack.

3. There was a shortage of reliable communications devices needed to establish lasting contact between the field artillery and the infantry. Thus the artillery support was often completely lacking at a decisive location. The infantry complained of insufficient help from its sister service.

4. The field artillery command stated: ". . . in order to be able to provide decisive action in infantry combat, (the field artillery must) dispense with the advantages of concealed positions and provide its fire mostly from open positions . . ." As shown in practice, in many cases this could not be done without paying the price of self-sacrifice.

5. In order to be able to place the field artillery guns where they were most urgently needed, the infantry requested their subordination.

6. The detaching of individual guns from the artillery units must naturally arouse the opposition of the artillery commander. In attack as well as in defense, the guns were valued by the effect of concentrated heavy fire. Any turned over to the infantry influenced the fulfillments of such tasks negatively.

7. Even individual platoons or batteries of the field artillery, with their guns and ammunition wagons, formed a very conspicuous target, prematurely betrayed troop concentrations for attack, and could be attacked on the battlefield by infantry weapons.

8. The detailing of field guns to the infantry, born of necessity, showed that even the Field Cannon 96 n.A., which had now become obsolete, was too big and too heavy to be moved for long distances by manpower. Transporting it by six-horse teams proved to be impossible.

9. In terms of the needs of the infantry, field cannons had an excessively great initial velocity of their shells, their trajectory was not variable enough . . . and the shot range was too great.

5 cm cannons in casemate mounts had been used as anti-assault guns in fortresses. Under the conditions of trench warfare the combat conditions were similar, for which reason the guns were taken out of the fortresses and moved to the front lines to support the infantry. This photo was taken in 1915.

The 3.7 cm antitank cannon (by Rheinmetall), on a rigid wheeled mount, weighed 175 kg and was thus practically ideal for the infantry. With a shot weight of only about 0.5 kilogram, though, the effect was insufficient, and besides, it could only be fired by direct aim.

The troops fighting in the eastern theater of war had already requested a particularly light gun in 1915. This was quite understandable, given the terrain and road conditions there. On the western front, the question of an infantry gun was not so urgent at first, and usable experience could be gained with the light 7.7 cm mine launcher and the 3.7 cm trench cannon. At the War Ministry in Berlin, though, the question of infantry guns had been pursued farther. It was clear that, on account of the tense situation in the realm of equipment, a new design could not be undertaken at that time. At the Krupp works, captured 7.62 cm guns had been reworked, and as of 1916 a total of ten batteries had been equipped with them. They were assigned to the assault battalions. One battery received the Austrian 3.7 cm Infantry Cannon M.15 for experimentation, and in the autumn of 1916 two additional batteries were set up with the 7.7 cm L/20 Infantry Gun made by Krupp. After the tank battle of Cambrai in 1917, upgraded requirements were made known: infantry guns also had to be suitable for antitank use. In order to save time, the F. Krupp AG developed the 7.7 c, Infantry Gun L/27, utilizing the barrel and other parts of the field cannon. It weighed 845 kg and fired 6.85 kg shells at distances up to 4600 meters. The initial velocity was 465 meters per second.

In the spring of 1918, eighteen newly established batteries were armed with the 7.7 cm Infantry Gun L/27. The Army High Command (OHL), though, halted the establishment of further batteries and the continuing mass production of infantry guns. Personnel and material were to be used to strengthen the field artillery. The attack battles in March and April 1918, though, showed how wrong this decision was. All too often, the infantry's assaults were brought to a standstill. Now the Army High Command demanded great numbers of infantry guns; even the 7.5 cm Mountain Cannon M.15 (by Skoda), which was scarcely suited to such use, was issued to the infantry.

The tactical experience gained from the use of infantry guns in 1918 was gathered in an experience report of the Infantry Instructional Regiment: "In waging offensive warfare, individual guns have proved to be not effective enough, so that every regiment of the attacking infantry was assigned a battery of light field artillery (7.5 cm cannons), the so-called infantry gun battery, also called an escort battery. It was under the command of the infantry commander, who usually deployed it so that it immediately followed the attacking infantry and fired on any resistance that appeared, particularly machine-gun nests or even still-intact enemy support points, in direct fire."

The light mine launcher could fire 7.7 cm mines with a weight of 4.46 kg up to 1050 meters. It weighed only 100 kilograms, but was only partially suitable for flat-trajectory use.

On the basis of wishes that now came from both the War Ministry and the Army High Command, the industry developed infantry guns. The designs of Rheinmetall attracted attention through their particularly low height of fire. Krupp introduced the 7.7 cm Infantry Gun 18, of which great numbers were ordered immediately.

The Infantry Gun 18 was a recoiling gun with a horizontal crank breech. The rear mount was equipped with an insertable traversing lever, which was supposed to make it possible to turn the gun quickly if tanks attacked from the flanks. The limber used was the uniform limber of the field artillery. The initial velocity of the shell, which weighed 6.85 kilograms, was some 350 meters per second, and it could be fired 5000 meters. The ammunition of the 7.7 cm field gun was used, though with a smaller propellant charge.

The infantry gun was officially supplied to infantry units as of May 1918. In all, 51 infantry gun batteries with 204 guns were established. In technical terms, the 7.7 cm Infantry Gun 18 was a milestone.

A 7.5 cm infantry gun in cratered terrain. Despite the low weight of some 300 kilograms, moving the gun was laborious and required the full strength of the crew. This photo was taken in the spring of 1918.

In 1918 the infantry possessed a total of three basically different escort weapons of artillery:

1. The 7.7 cm infantry gun, used to attack targets in direct fire.

2. The 7.7 cm mine launcher, used to attack targets in indirect fire.

3. The 3.7 cm antitank cannon, used to attack armored targets in direct fire.

These were versatile and necessary weapons. In terms of organization and logistics, they were naturally also a source of problems for the infantry. Thus it is quite understandable that many infantry officers suggested that in the postwar era a support artillery weapon be developed to handle the three tasks noted above. This was no great problem for the first two tasks. Only the repeated efforts to develop an effective antitank weapon simultaneously burdened the theoretical discussion and practical developmental work in this area after the war.

Outside Germany, theoretical discussion bore fruit very soon in the form of many different designs for infantry guns. In Germany, discussion had to be limited at first to a lively and controversial discussion, carried on by, among others, the "Militärwochenblatt" and the "Artilleristischen Rundschau." Among the suggested solutions were guns with howitzer and antitank barrels beside, over or under each other. Only the essential recognition that the two basically different tasks of providing antitank defense and infantry support could be met satisfactorily only with two different gun designs opened the way for the development of infantry guns. In Germany, this decision, surely encouraged by the many practical experiences gained in the war, had been made already by the early twenties. In 1927 the development of a device that, for reasons of secrecy, was called a "Light Mine Launcher 18", was concluded. At the same time, the continuing theoretical discussion had the effect outside Germany, where all German military activity was looked upon with suspicion, of concealing knowledge of the level of development that had actually been attained.

Not very popular as an infantry escort gun — the Skoda 7.5 cm mountain cannon. In the autumn of 1916, two batteries were supplied with this special gun. In 1928 the Reichswehr still possessed 23 of the 7.5 cm Skoda M.15 Mountain Cannon.

A multipurpose gun made by Skoda. It weighed 157.5 kg and had a caliber of 70 mm (shot weight 3 kg). For antitank use, a 19 kg barrel type with a caliber of 32 mm could be installed (shot weight 0.5 kg).

The development of the Light Mine Launcher 18, the later 7.5 cm Infantry Gun 18, had been based on the following general requirements:

1. The gun had to be light and mobile, capable of following the infantry in combat terrain pulled by manpower.

2. It had to offer the enemy a small target, and thus have a low firing height.

3. It was also important that relatively small targets at close and medium ranges could be attacked in direct fire with large propellant charges, and in indirect fire with light charges, with equal success.

4. Sufficient effect of single shells was required; it was made clear that the caliber was not to be less than 75 mm.

5. In addition to being moved by manpower, there was also to be the possibility of being towed for great distances behind a limber by horsepower. Only in the following years, with the growing motorization of the infantry, was towing by motor vehicles considered.

At the Rheinmetall-Borsig AG, these military requirements resulted in a howitzer with a field of elevation from 0 to 75 degrees and a caliber of 75 mm, a shell of this size promising a satisfactory effect on a target. A propellant charge of, at first, five parts, also made it possible to strike targets behind cover at short ranges. Because of the con-

siderably lighter weight, a box mount had been selected. The disadvantage was that a small field of traverse had to be accepted. The gun barrel was noteworthy: It was mounted in a tipping gun cradle.

The gun was aimed by using a visible aiming device with an independent sight line and a parabolic scope, which gave fourfold magnification of a ten-degree field of vision. A five-piece gun shield, 3 to 4.5 mm thick, was to offer the gun crew protection and cover.

As already noted, this infantry gun was ready to be introduced in 1927. In the Wehrmacht Procurement Schedule No. 361/29 (Secret Command Material) of March 25, 1929, it was planned to purchase 148 Light Mine Launcher 18 units for the infantry between 1928 and 1932, at a cost of 2,738,000 Reichsmark. It was expressly noted that at this point in time there were already 14 devices of this type on hand. In addition, 127,200 light explosive mines were to be ordered.

On July 5, 1930, the Wehrmacht authorized the production of additional Light Mine Launcher 18 units, including their equipment and limbers. The money to be used for them was obtained by canceling the production of 38 Medium Mine Launcher 16/18 units. According to a planning schedule of 1928, the available number of the Light Mine Launcher 18 should assure the supplying of 16 divisions of the Reichswehr.

Infantry guns being towed by bicycles. In Switzerland, the staff companies of the infantry battalions used the 47 mm M.31 Cannon as both an antitank and infantry gun. The explosive shell weighed 1.75 kilograms.

In Belgium too, am antitank gun with a caliber of 47 mm was preferred, so that it could also be used for infantry support thanks to the greater weight of the explosive shell, which weighed 1.55 kilograms.

On September 10, 1932, the H.Dv. 105/2 "The Mine Launcher, Part 2: Description of the Light Mine Launcher 18" was published. Only in April 1937 did the additional pages 39-46 appear, in which the weapon was renamed Light Infantry Gun 18. On September 1, 1939, 2933 light 7.5 cm infantry guns were in the possession of the Wehrmacht. The production cost was 6700 Reichsmark (in 1936 it had been 8300 Reichsmark).

The Light 7.5 cm Mountain Infantry Gun 18 had been introduced in 1937. In view of the needs of the mountain troops, which needed guns with a large field of traverse (here 50 instead of 11 degrees) for use in limited space in the mountains, where a short spreading mount could also be advantageous, Rheinmetall had developed such a gun. The spars of the spread mount could be shortened, and transporting in five to seven pack-horse or human loads was also possible. Otherwise it corresponded to the Light 7.5 cm Infantry Gun 18.

The Rheinmetall-Borsig AG also offered an alternative proposal for a mountain infantry gun. Instead of the tipping-barrel breech, it had a horizontal crank-operated breech.

Light infantry guns, none of them larger than 75 mm in caliber, were also developed in other countries. In Germany, World War I experience always raised doubts as to whether light 7.5 cm infantry guns were really capable of fighting down all the opposition that might face the infantry. To be able to shoot down barricades, heavily covered field fortifications and buildings expanded into nests of opposition, the effect of the 15 cm field howitzer was needed. It was believed that, in the interests of saving weight, a lesser shot range could be lived with.

Developmental work on a heavy 15 cm infantry gun began in 1927 under the code name of "Medium Mine Launcher." There were essential differences from the medium 17 cm mine launcher, which had been used by the Reichswehr since World War I. The new gun was a rear loader and had a box mount equipped with wheels.

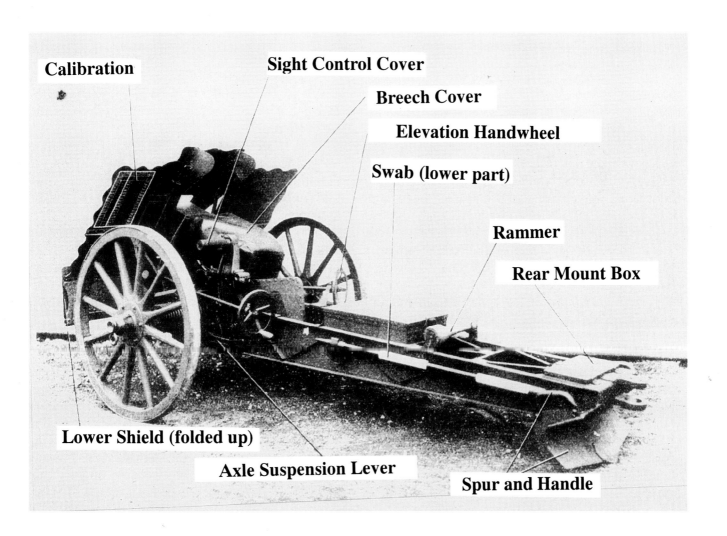

Calibration   **Sight Control Cover**   **Breech Cover**   **Elevation Handwheel**   **Swab (lower part)**   **Rammer**   **Rear Mount Box**   **Lower Shield (folded up)**   **Axle Suspension Lever**   **Spur and Handle**

The Light Mine Launcher 18 (later the Light 7.5 cm Infantry Gun 18).

The medium mine launcher (later the heavy 15 cm Infantry Gun 33).

In the medium mine launcher too, the use of a box mount saved weight. Launching the shell, which weighed more than 1.5 tons, in a new direction of traverse proved to be problematic. Alternative solutions were sought. The heavy 15 cm infantry gun was developed and tested with a spread mount with a traverse of 40 degrees. The weight climbed to 2100 kilograms, though the shot weight had been lowered to 38 kilograms. In addition, Rheinmetall had suggested infantry guns with box and spread mounts. These guns, with a caliber of 10.5 cm, promised a shot effect that was comparable to that of the light field howitzer.

The high weight of the heavy 15 cm infantry gun with a spread mount and the comparatively meager shot effect of the 10.5 cm infantry gun (13.5 kg shot mass instead of 14 kg) decided the matter in favor of the design with the box mount. This took place in 1933. Only as of 1936-37-, though, did the heavy infantry gun reach the troops in noteworthy numbers. In 1937 there appeared the H.Dv. 130/4a "The Infantry Gun Company", which provided the rules for organizing and utilizing the 13th (Infantry Gun) Companies in the infantry regiments. Only on June 15, 1938 was H.Dv. 109 "The Heavy Infantry Gun" issued — and the firing instructions for the infantry gun companies appeared even later (July 1939).

On September 1, 1939, there were 410 of the heavy 15 cm Infantry Gun 33 in service with the troops. The price for one unit was 20,450 Reichsmark.

The relatively great weight of the heavy infantry gun led, in the summer of 1939, to attempts to mount it on tank chassis and thus turn it into a mobile support weapon for the riflemen in the Panzer divisions.

The Heavy Infantry Gun 33 mounted on Gun Vehicle 38 (equipment numbers 5-1523 and 5-1532), of which well over 400 were produced by November 1944, proved itself very well. They were utilized in the heavy infantry gun companies of the Panzer grenadier regiments.

Equipping divisions with infantry guns was varied and changed numerous times during the course of the war. At the beginning of World War II, an infantry division had a total of eighteen light and six heavy infantry guns in its three rifle regiments. In 1942 the 2nd SS Panzer Grenadier Regiment "Das Reich" was supposed to have twenty light and eight heavy infantry guns. In the 78th Assault Division there were fourteen light and six heavy infantry guns in 1943. In the planned wartime strength of an infantry division in 1945 there were to be 29 of the light and six of the heavy infantry guns.

Right: Alternative suggestions: The 10.5 cm infantry gun on a spread mount with a folding spur. In firing position it weighed 1110 kilograms.

Below: The heavy 15cm infantry gun on a spread mount. It weighed 2100 kilograms and could fire shells weighing 38 kilograms as far as 5000 meters.

Below: A design drawing of a 10.5 cm infantry gun with a box mount. A weight of 1040 kilograms in firing position was planned.

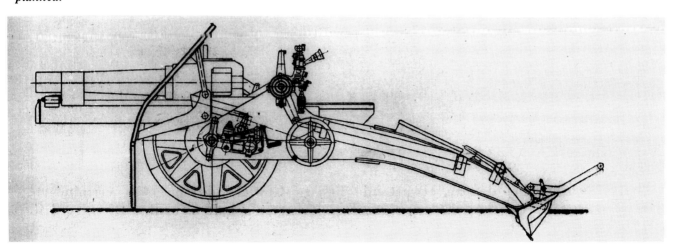

In general it can be estimated that the infantry gun proved itself well, especially where it, thanks to its mobility and firepower, could advance the forward movement of the attacking infantry. Organization, formation and proper training contributed to providing the infantry with the necessary superiority at focal points. But naturally there were also disadvantages to such a weapon designed for very special uses. All infantry guns were only of limited use as antitank weapons, a task which took on decisive importance in the war and had to be carried out by all types of service arms.

On July 1, 1940, as ordered, the 7.5 cm Infantry Gun Shell 38 HL/A, an armor-piercing projectile, was introduced for the 7.5 cm infantry gun. With this shell, regardless of the range, 75 millimeters of armor plate could be pierced (later, with the HL/B version, this was raised to 90 mm). The comparable ammunition of the heavy 15 cm infantry gun even had a penetrating power of 140 to 160 mm. In spite of that, the infantry guns' chances of success in antitank action were limited. The shot range in direct fire and the field of traverse were simply too limited.

Thus it is no surprise that the suggestion of the Infantry Inspection appeared in the work records of the Army Weapons Office/Weapon Testing Section I, that the penetrating power of the infantry gun ammunition be upgraded over 75 mm. The primary combat range against tanks was to be extended to 2000 meters.

A further disadvantage was the meager shot range of all infantry guns. The light 7.5 cm Infantry Gun 18 attained 3500 meters, or 4600 meters with a special cartridge, and the heavy 15 cm infantry gun could fire a maximum of 4700 meters. If targets that were outside the range of the infantry guns were to be engaged, other weapons had to be brought in. One example: The 384th Infantry Division, in the sector of Infantry Regiment 535, lost its last two 5 cm antitank guns on October 20, 1942. They were detailed off for use against Russian field cannon, by which they, along with their crews, were destroyed within a few minutes.

During the course of World War II, a considerable new increase in the effect of weapons on the battlefield became evident. There was little room left for expensive special weapons such as the German infantry guns. Therefore in the Speer Program of 1943-1944, the viewpoint was expressed that, in the interests of rationalization, the light 7.5 cm infantry gun should be taken out of production. One part of its tasks should be taken over by the 8 cm Medium Grenade Launcher 34. As a stopgap mea-

An alternative suggestion for the light 7.5 cm infantry gun made by Rheinmetall. The gun, weighing 426 kilograms, had a crank breech, and the spars could be shortened.

sure, a 7.5 cm gun should be mounted temporarily on the mount of the 3.7 cm antitank gun. For the heavy 15 cm infantry gun, it was planned to increase production to 150 units per month. Whether the production of this gun should also be halted would depend on the progress made in developing a heavy 15 cm grenade launcher.

There was no shortage of attempts to upgrade the effectiveness of the infantry guns themselves. As early as 1940, the prototype of a Light 7.5 cm Infantry Gun 42 had been introduced by Krupp, and a version of it with a smooth bore appeared in 1944.

Falling short of meeting both demands — for better antitank effectiveness and longer range — was an infantry gun that had been introduced originally as the Panzerjäger Cannon 37 and, as of June 15, 1944, borne the designation of 7.5 cm Infantry Gun 37. A makeshift solution, in which the 7.5 cm L/24 barrels were mounted on the still available, superfluous mounts of the 3.7 cm Antitank Cannon L/24, was tried. The effective shot range against tanks was about 500 meters; otherwise targets to a distance of 5150 meters could be fired on. Between May and December 1944, 2278 units were produced.

There followed the 7.5 cm Infantry Gun 42, 527 of which were delivered between September 1944 and March 1945. It differed from the Infantry Gun 37 in its mount, which had been taken over from the 8 cm Panzerwurfkanone 600 (8 H 63).

In the "Assault Program of the Army" (OKH, GenSt d H Abt. III Gruppe Planung No. 8767/45 gK.Chefs.) of January 9, 1945, the cessation of production of the 18, 37 and 42 Light Infantry Guns was envisioned. Production of the Heavy 15 cm Infantry Gun 33 was also decreased by 200 units in December 1944, leaving it too with no great hopes for the future. In part, heavy grenade launchers were supposed to take over their tasks.

An interesting facet in the arming of the infantry with infantry guns appears in a note about a conference held by General Buhle, the chief of Army Armaments, on February 24, 1945. In the course of strengthening the equipment of the division artillery with 7.5 cm field guns, it was planned to introduce the 8 cm Panzerwurfkanone 8 H 63 (previously PWK 600) in the infantry regiments. This new development could be utilized as both an infantry gun and an antitank weapon, and weighed only 600 kilograms. For antitank use, hollow launcher grenades (4462) were fired up to 750 meters. Their penetrating power amounted to 140 mm of armor plate. For explosive shrapnel shells, the Launcher Grenade 5071 could be fired up to 6200 meters. This ammunition was stabilized in flight. Test models of this weapon, made by Rheinmetall and Krupp, had been introduced and tested at the Unterlüss firing range in September 1944. There the prototype from the Rheinmetall-Borsig AG gave better results. At that time, when the Army High Command was formulating an assault program, 44 of the Panzerwurfkanone were at various fronts for troop testing, as well as in Döberitz and Mittenwalde.

These antitank launchers were supposed to take over the tasks of the earlier infantry guns and infantry antitank guns at the regimental level. Technically, there was no comparison between the two weapons. The same was true of the heavy infantry guns, whose tasks were to be taken over in part by the heavy grenade launchers. Naturally, the decisions on the infantry gun question in the spring of 1945 had to be made above all on the basis of the complicated situation of the armament industry in Germany. But that was the end of the history of a specialized kind of artillery, the production, organization and utilization concept of which were attuned to the infantry. In Germany it had been developed in this particular fashion.

A typical shooting procedure practiced and used by the infantry gun companies was the ricochet shot.

Schützenmulde

STRUCTURE OF AN
INFANTRY REGIMENT (1939)

Regimental staff

Staff Company

1st Infantry Battalion
3 Infantry Companies (1st-3rd)
1 Machine Gun Company (4th)

2nd Infantry Battalion
3 Infantry Companies (5th-7th)
1 Machine Gun Company (8th)

3rd Infantry Battalion
3 Infantry Companies (9th-11th)
1 Machine Gun Company (12th)

13th (Infantry Gun) Company
Company Troop

1st (light) Platoon with 2 light
7.5 cm Infantry Gun 18

2nd (light) Platoon with 2 light
7.5 cm Infantry Gun 18

3rd (light) Platoon with 2 light
7.5 cm Infantry Gun 18

4th (heavy) Platoon with 2 heavy
15 cm Infantry Gun 33

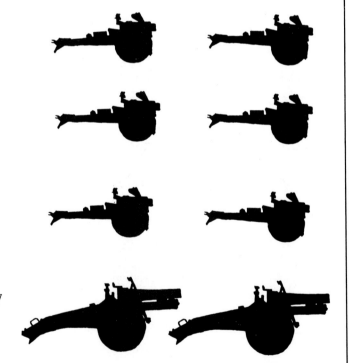

14th (Infantry Panzerjäger) Company

Baggage and Supply Troops

LIGHT 7.5 cm
INFANTRY GUN 18
(MOTORIZED AND HORSEDRAWN)

Caliber:	75 mm
Barrel length:	835 mm
	= L/11.8
Initial velocity:	221 m/s
with special charge:	260 m/s
Shot range:	3550 m
with special charge:	4600 m
Traverse range:	11º
Elevation range:	-10º to +75º
Weight ready to fire:	
motorized:	570 Kg
horsedrawn:	400 Kg
Weight ready to march:	
motorized:	515 Kg
horsedrawn:	405 Kg
Shot weight:	
Infantry Shell 18	5.45 Kg
Infantry Shell 38 HL/A	3.1 Kg
Infantry Shell 38 HL/B	3.5 Kg
Rate of fire:	8-12 rounds/ minute
Penetrating power:	
Infantry Shell 38 HL/A	75 mm armor plate
Infantry Shell 38 HL/B	90 mm armor plate

Manufacturers: Böhmische Waffenfabrik, Strakonitz; HABÄMFA, Ammendorf-Halle

Above: Action in the ruins of Stalingrad, early November 1942.

Below: Light 7.5 cm Infantry Gun 18 (horsedrawn) in firing position. An advantage in many battles, noticeable here, was the low profile. The crew was protected by a steel shield.

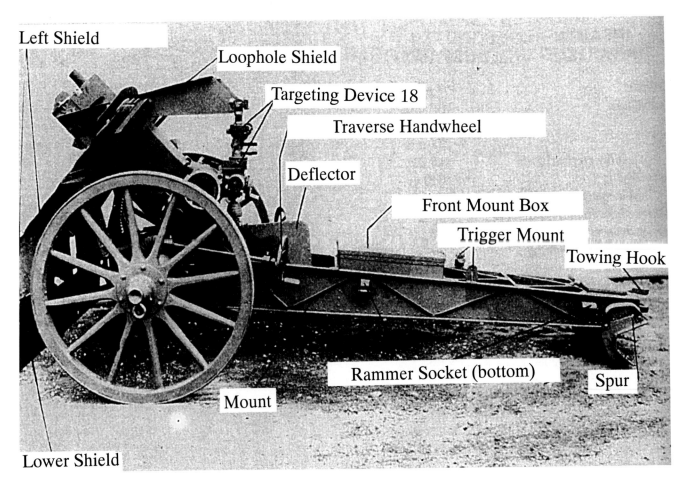

Left Shield

Loophole Shield

Targeting Device 18

Traverse Handwheel

Deflector

Front Mount Box

Trigger Mount

Towing Hook

Rammer Socket (bottom)

Spur

Mount

Lower Shield

The 7.5 cm Infantry Gun in firing position.

Training of a 7.5 cm infantry gun crew. The gun is painted in the typical three-color camouflage of the 1930s.

Above and below: Training in aiming the gun. The gun crews were taught to aim the gun quickly and accurately, a prerequisite for quick and effective firing (H.Dv. 111/2 "Firing Instructions for Infantry Guns", 1939).

Light 7.5 cm infantry guns were towed behind the limber and ammunition carrier (Itf. 14) by four-horse hitches.

18   *Firing at a high angle, with raised aiming gear. France, June 1940.*

Above: Training in prewar times. Note the three-color camouflage paint.

Below: Action in the winter of 1941-1942. The front of the shield and the wheels are camouflaged with white paint.

Front and rear views of the light 7.5 cm Infantry Gun 18 (horsedrawn) at the Military Technology Study Collection in Koblenz.

Above: The light 7.5 cm Infantry Gun 18 (horsedrawn) in the Museum of Military History in Dresden. The lower shield is folded up.

Below: The version of the light 7.5 cm Infantry Gun 18 made for motorized towing, in the Military Technology Study Collection in Koblenz.

This infantry gun is being pulled to its firing position by its crew. On the mount are baskets of ammunition. This photo was taken in France in June 1940.

Below: Training replacement infantry gun crews with the Infantry Gun 18 (mot) in the Cottbus area, summer 1942.

Left: A Light Infantry Gun 18 (mot) in a hastily prepared firing position on the eastern front in 1942.

Below: Opening fire from an open firing position on the eastern front in 1942.

Both sides: Action photos of light infantry guns of a motorized advance guard in the streets of a small Russian city, July 1942.

The Light 7.5 cm Infantry Gun 18 (mot) on tow behind an Opel 1.5 ton truck used as a troop carrier as well as a towing vehicle in Poland, September 1939.

Below: The rugged Krupp L2 H143 towing truck (Kfz. 69) often was used to tow the Light 7.5 cm Infantry Gun 18, as here on the eastern front in the spring of 1942.

Above: Here the Medium Uniform PKW (Kfz. 15) is used to tow the 7.5 cm Infantry Gun 18 and the single-axle ammunition trailer on a training mission in the summer of 1939.

Below: The Medium Uniform PKW (Kfz. 15) towing an infantry gun, and another of the same kind towing an ammunition trailer in France, June 1940.

The firing position for a Light 7.5 cm Infantry Gun 18 (from Instruction Book 57/5 "Illustrated Modern Position Building" of June 1, 1944.

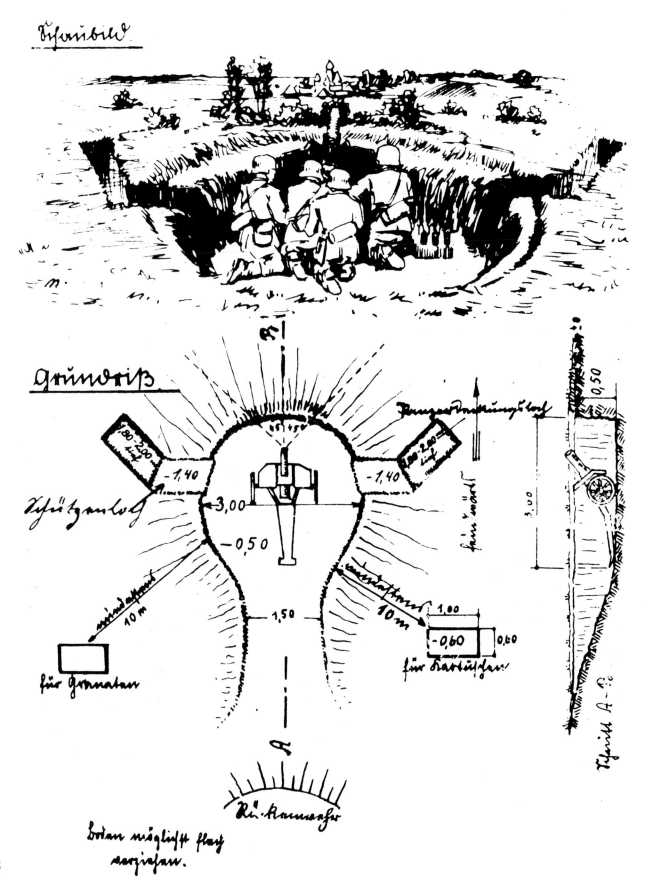

LIGHT 7.5 CM MOUNTAIN INFANTRY GUN 18

Caliber:	75 mm
Barrel length:	835 mm
	= L/11.8
Initial velocity:	221 m/s
Shot range:	3550 m
Traverse field:	35º
Elevation field:	-10º to +75º
Weight ready to fire:	440 Kg
Weight ready to march	410 Kg
Shot weight:	
Infantry Shell 18	5.45 Kg
Infantry Shell 38 HL/A	3.1 Kg
Infantry Shell 38 HL/B	3.5 Kg
Rate of fire:	8-12 rounds/
	minute
Penetrating power:	
Infantry Shell 38 HL/A:	75 mm
	armor plate
Infantry Shell 38 HL/B	90 mm
	armor plate

Manufacturers: Böhmische Waffenfabrik,
Strakonitz; HABÄMFA, Ammendorf-Halle

The light 7.5 cm Mountain Howitzer —an alternative suggestion from Rheinmetall for a light 7.5 cm mountain infantry gun. The spars are shortened.

A 7.5 cm Mountain Infantry Gun 18 in firing position at Kuban, spring 1943. Here the shield has been removed.

Above and below: Training with the light 7.5 cm Mountain Infantry Gun in Mittenwalde, winter 1939-1940.

The gun could be dismantled quickly and made into six donkey loads or ten loads for the crew. Some of these guns were also used by paratroopers.

Ammunition for the l. u. G. 18
7.5 cm Jgr. 18

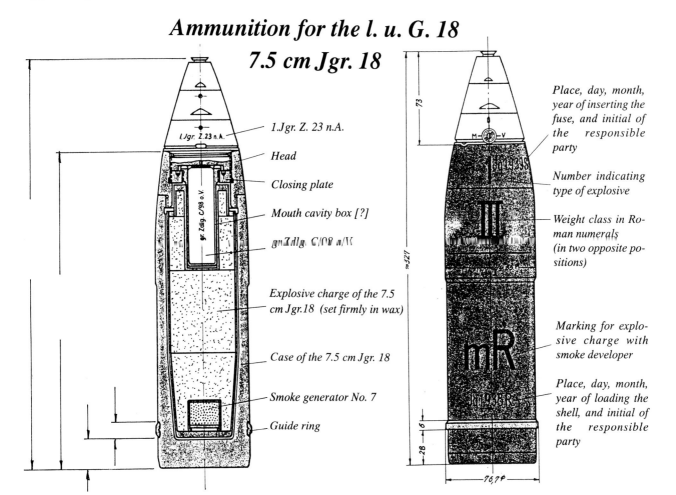

1.Jgr. Z. 23 n.A.

Head

Closing plate

Mouth cavity box [?]

Explosive charge of the 7.5 cm Jgr.18 (set firmly in wax)

Case of the 7.5 cm Jgr. 18

Smoke generator No. 7

Guide ring

Place, day, month, year of inserting the fuse, and initial of the responsible party

Number indicating type of explosive

Weight class in Roman numerals (in two opposite positions)

Marking for explosive charge with smoke developer

Place, day, month, year of loading the shell, and initial of the responsible party

7.5 cm Jgr. 18 with Smoke Generator No. 7

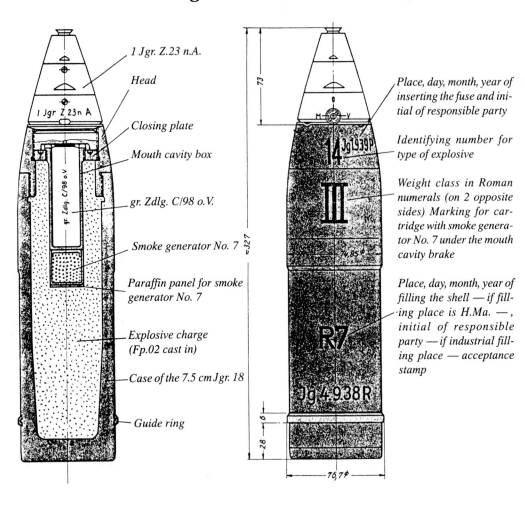

1 Jgr. Z.23 n.A.

Head

Closing plate

Mouth cavity box

gr. Zdlg. C/98 o.V.

Smoke generator No. 7

Paraffin panel for smoke generator No. 7

Explosive charge (Fp.02 cast in)

Case of the 7.5 cm Jgr. 18

Guide ring

Place, day, month, year of inserting the fuse and initial of responsible party

Identifying number for type of explosive

Weight class in Roman numerals (on 2 opposite sides) Marking for cartridge with smoke generator No. 7 under the mouth cavity brake

Place, day, month, year of filling the shell — if filling place is H.Ma. —, initial of responsible party — if industrial filling place — acceptance stamp

Ammunition with a smoke generator allowed quicker targeting of a known target. With the 7.5 cm Infantry Shell 38, armored vehicles could be attacked with hope of success. The penetrating power was 75 mm; the effective range was under 500 meters. With a lot of luck, armored targets could be hit at ranges up to 1500 meters.

7.5 cm Jgr. 38

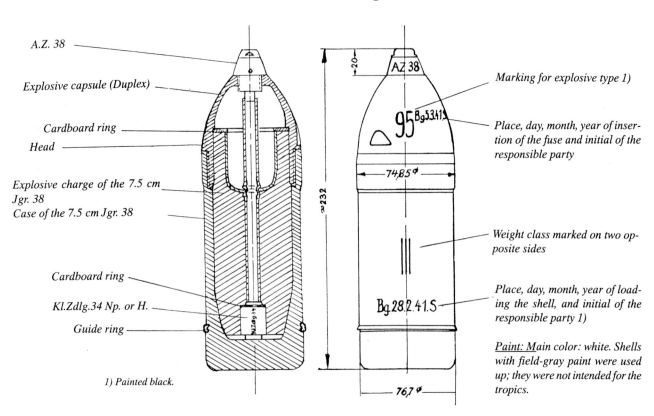

A.Z. 38

Explosive capsule (Duplex)

Cardboard ring

Head

Explosive charge of the 7.5 cm Jgr. 38

Case of the 7.5 cm Jgr. 38

Cardboard ring

Kl.Zdlg.34 Np. or H.

Guide ring

1) Painted black.

Marking for explosive type 1)

Place, day, month, year of insertion of the fuse and initial of the responsible party

Weight class marked on two opposite sides

Place, day, month, year of loading the shell, and initial of the responsible party 1)

Paint: Main color: white. Shells with field-gray paint were used up; they were not intended for the tropics.

Shell cartridge of the I.J.G. 18

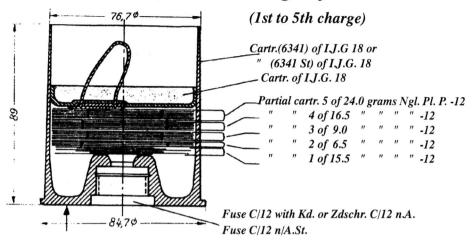

(1st to 5th charge)

Cartr.(6341) of I.J.G 18 or
" (6341 St) of I.J.G. 18
Cartr. of I.J.G. 18

Partial cartr. 5 of 24.0 grams Ngl. Pl. P. -12
" " 4 of 16.5 " " " " -12
" " 3 of 9.0 " " " " -12
" " 2 of 6.5 " " " " -12
" " 1 of 15.5 " " " " -12

Fuse C/12 with Kd. or Zdschr. C/12 n.A.
Fuse C/12 n/A.St.

76,7⌀   84,7⌀   89

7.5 cm Infantry
Shell 39 with 38
steel impact fuse

Portable container with three special fuses made of steel for the 7.5 cm Infantry Gun 18.

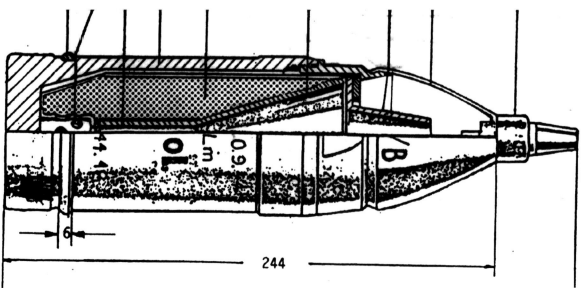

Left: Originally the shell cases were made of brass. Then were increasingly replaced by steel cases.

6   244

Above: Guns captured from the enemy were very often used as infantry guns. Among the rarities is this Russian 4.5 cm Antitank Cannon 184/6(r)m which was used by the Germans in the northern sector of the eastern front in the autumn of 1941.

Right: Captured in great numbers and used by the Wehrmacht until the war's end was the Russian 7.62 cm Infantry Cannon-howitzer 290(r) 27.

Left: The gun could fire explosive shells weighing 6.4 kilograms up to a distance of 8550 meters. In firing position, though, it weighed 780 kilograms.

HEAVY 15 cm
INFANTRY GUN 33
(MOTORIZED AND HORSEDRAWN)

Caliber:	149.1 mm
Barrel length:	1700 mm
Initial velocity:	
Infantry Shell 38:	240 m/s
Infantry Shell 39HL/A:	280 m/s
Shot range:	4700 m
Traverse field:	11º
Elevation field:	-4º to +75º
Weight ready to fire:	
Motorized:	1800 Kg
Horsedrawn:	1680 Kg
Weight ready to march:	
Motorized:	1825 Kg
Horsedrawn:	1700 Kg
Shot weight:	
Infantry Shell 38:	38 Kg
Infantry Shell 39 HL/A:	24.6 Kg
Rate of fire:	2-3 rounds/ minute
Penetrating power:	160 mm armor plate

Manufacturers: AEG, Henningsdorf Works;
Böhmische Waffenfabrik, Strakonitz

Above right: Mid-April 1941: A heavy 15 cm infantry Gun of the LAH Division fires on the approaches to the Klisura Pass.

Below: A 15 cm Infantry Gun 33 (horsedrawn) on the march along ever-muddier Russian roads on the eastern front, October 1941.

Prewar photos, published in the journal "Die Wehrmacht", showing series production of the Heavy 15 cm Infantry Gun 33. In the front row are the Heavy 15 cm Field Howitzer 18.

A 15 cm infantry gun in firing position in a cornfield on the eastern front, southern sector, summer 1942.

Below: On the return march near Rouen, summer 1944. The Medium 8-ton Towing Tractor (Sd. Kfz. 7) presumably is being used here only temporarily to tow the 15 cm Heavy Infantry Gun 33 (mot). Structurally, the one-ton towing tractor (Sd.Kfz. 10) was used to tow it.

The heavy infantry gun in firing position behind walls of snow on the eastern front, winter 1941-1942.

Below: In action with the 14th Panzer Division on the eastern front, spring 1942.

With the Heavy 15 cm Infantry Gun 33 —here the type for motorized towing —more heavily covered targets were to be knocked out quickly on the battlefield. Earth cover up to two meters thick could be penetrated by using delayed fuses; about 1000 splinters resulted when the 38-kilogram heavy shell exploded.

Below: An accident on the road. The gun crew has already set up the heavy infantry gun again and is now trying to turn the one-ton towing tractor (Sd.Kfz.10) right side up. Eastern front, winter 1943-1944.

Firing Position for s.J.-G.
with tank cover and ammunition holes

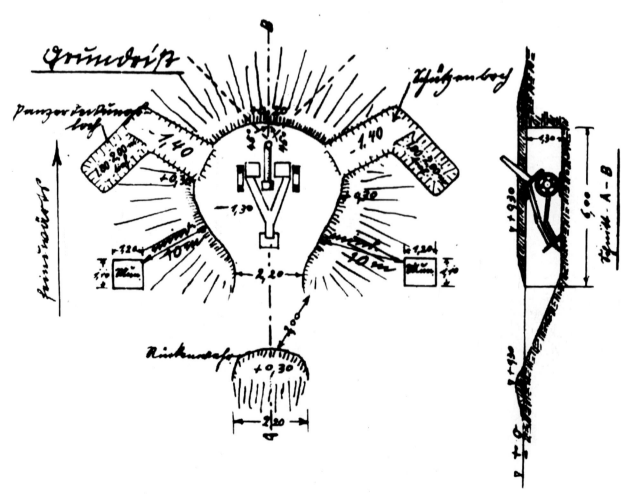

A 15 cm Infantry Gun 33 (horsedrawn) captured by the Red Army on the eastern front, December 1941.

Below: The war's end in the Erzgebirge, 1945. Among the plentiful scrap that was assembled at the Geising and Altenberg railroad stations after the war ended are two Heavy 15 cm Infantry Gun 33 units. At right is the final version.

The 15 cm Infantry Gun 33 (horsedrawn) at the Museum of Military History in Dresden.

The 15 cm Infantry Gun 33 outside the former DDR Army Museum in Potsdam. The gun has iron-tired spoked wooden wheels and a simplified straight shield.

A tank chassis, an infantry gun, and a few armor plates, and the self-propelled gun is finished. The work on a 15 cm sIG 33 (Sf.) is ready for the 9th Panzer Division in the Balkans, spring 1941.

Below: In the summer of 1940, the Army had six heavy infantry gun companies (Sf.) (No. 701-706), which were assigned to Panzer divisions. All were equipped with 15 cm sIG 33 (Sf.) on Panzer I, Type B chassis. The last vehicles of this type were with the 5th Panzer Division in Russia in mid-1943.

Advancing in the Balkans, April 1941. Naturally, the 8.5-ton vehicles were overburdened and, being 2.8 meters high, impossible not to see. The self-propelled guns on Panzer I chassis gained experience that was evaluated for later designs.

Unloading a Heavy 15 cm Infantry Gun 33 on Panzer I chassis in the harbor of Tripoli. Twelve vehicles of this kind were built; they were used by two heavy infantry gun companies (Sf.) (No. 707 and 708) of the German Afrika-Korps.

The best results were gained with the Heavy 15 cm Infantry Gun 33 (Sf.) on Panzer 38 (t), Type H chassis (Sd.Kfz. 138/1), which were produced as of February 1943 and used by the heavy infantry gun companies of the Panzergrenadier regiments.

In firing position on the eastern front, spring 1944.

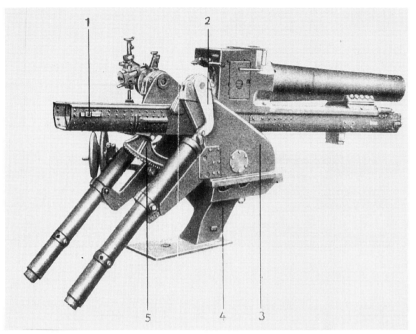

Left and below: the Heavy 15 cm Infantry Gun 33, dismantled.

Left: A look at the fighting compartment. The racks at right and the box on the engine compartment carried the 15 cm shells. At right, ahead of the loader's seat, are containers for the shell cases.

LIGHT 7.5 cm INFANTRY GUN 37

Caliber:	75 mm
Barrel length:	1798 mm
	= L/24
Initial velocity:	80 m/s
Shot range:	5150 m
Traverse field:	58º
Elevation field:	-10º to +40º
Weight ready to fire:	510 Kg
Weight ready to march:	— -
Shot weight:	
Infantry Shell 18:	5.45 Kg
Infantry Shell 38 HL/A:	3.5 Kg
Rate of fire:	— -
Penetrating power:	
Infantry Shell 38 HL/B:	90 mm
	armor plate

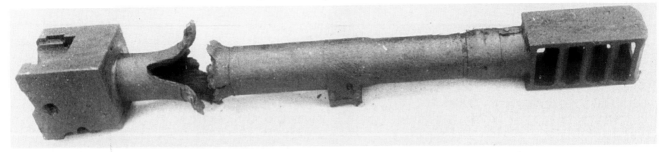

Above: The barrel and breech of a Light 7.5 cm Infantry Gun 37, ruined in classic style by a barrel explosion. Found in Halbe, owned by the Museum of Military History in Dresden.

Below: This Light 7.5 cm Infantry Gun 37 belongs to the Museum of Handicrafts, Natural History and Weapons Technology in Dahn-Reichenbach. In September 1944 there were 1211 of them.